FOCUS ON
WEATHER
AND CLIMATE

BARBARA TAYLOR

GLOUCESTER PRESS
London · New York · Sydney

© Aladdin Books Ltd 1993

Designed and produced by
Aladdin Books Ltd
28 Percy Street
London W1P 9FF

First published in
Great Britain in 1993 by
Watts Books
96 Leonard Street
London EC2A 4RH

ISBN 0 7496 1324 6

A catalogue record for this book is
available from the British Library.

Printed in Belgium

Design	David West Children's Book Design
Designer	Flick Killerby
Series director	Bibby Whittaker
Editor	Suzanne Melia
Picture research	Emma Krikler
Illustrator	David Burroughs

The author, Barbara Taylor has a degree in
science, and has written and edited many
books for children, mainly on science
subjects.

The consultant, Martin Weitz is a journalist
at the BBC. He specialises in health and
the environment.

INTRODUCTION

The ever-changing weather is an important part of daily lives all over the world. And until a hurricane strikes, it is easy to forget just how powerful the weather can be. People now understand a great deal about how the Sun causes the weather, and how the climate varies from the Poles to the Equator. With the help of computers, we try to forecast changes in the weather, but we are only just beginning to understand how pollution is changing the weather and climate of the whole planet. This book explores the scientific principles behind the weather, and aims to provide a complete picture of weather and climate in the past, present and the future.

Geography
The symbol of the planet Earth indicates where geographical facts and activities are examined in this book. These sections include a discussion of why sea breezes occur and how world climates effect skin colour.

Language and literature
An open book is the sign for activities and information about language and literature. These sections include a look at Zeus the Greek god of thunder and discusses sayings and rhymes about the weather.

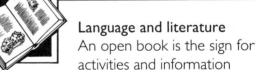

Science and technology
The microscope symbol shows where science information or activities are included. One of these sections discusses how to identify animals by their footprints.

History
The sign of the scroll and hourglass indicates historical information. These sections examine how the Romans heated their houses and looks at the history of weather forecasting. The history of windmills is also discussed in one box.

Social history
The symbol of the family indicates where information about social history is given. One box takes a look at ancient civilisations who worshipped the Sun as a god, and another discusses seasonal festivals.

Art, craft and music
The symbol showing a sheet of music and art tools, signals where activities and information about art, craft and music are given. The symbolic use of clouds in religious paintings is the subject of one section.

CONTENTS

WHAT IS THE WEATHER? 4/5

EXTREME CLIMATES............................6/7

TEMPERATE CLIMATES 8/9

SEASONS 10/11

LIVING WITH THE WEATHER 12/13

SUN & TEMPERATURE 14/15

WINDY WEATHER 16/17

CLOUDS 18/19

RAINY DAYS.....................................20/21

SNOW, ICE & HAIL 22/23

EXTREME WEATHER.........................24/25

CHANGING THE WEATHER..............26/27

FORECASTING...................................28/29

WONDERFUL WEATHER FACTS............30

GLOSSARY 31

INDEX...32

WHAT IS THE WEATHER?

From sunshine and showers to blizzards and hurricanes, the weather is a combination of wind, rain, clouds and temperature. Believe it or not, all our weather is caused by the air around our planet warming up and cooling down. The average weather in one particular region is called the climate. In some climates, the weather stays much the same all year round. But in many parts of the world, the weather changes at certain times of year. A climate apears to stay the same, but may change quite a lot over thousands of years.

Earth moves around the Sun once a year

Earth spins once a day

The atmosphere

The Earth is surrounded by a thick blanket of air called the atmosphere, which is made up of five layers. Weather happens only in the layer nearest to the Earth – the troposphere. This stretches up about 11km above the surface of the planet, not much higher than the top of Mount Everest. The troposphere is the warmest layer of the atmosphere and contains the most moisture.

Auroras are produced when radiation from the Sun hits the outer layers of the atmosphere.

80 km

50 km

The Sun and Earth

The Earth moves slowly around the Sun once every year. Because the Earth is tilted, places are closer to the Sun at different times of year. This affects the amount of light and heat these places receive, and produces a pattern of changes in the weather called the seasons. The Earth also spins on its axis once every 24 hours, giving us night and day.

Weather occurs in this layer, the troposphere.

11 km

Sun gods

Many ancient peoples worshipped the Sun as a god. They made sacrifices to the gods to keep the Sun shining. In the Aztec religion, the Sun was the warrior, Huitzilopochtli, who died every evening to be born again the next day, driving away the stars and Moon with a shaft of light.

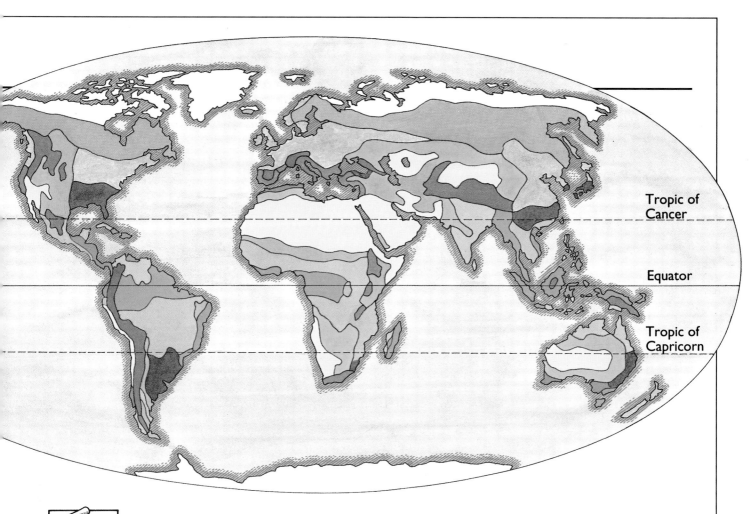

Tropic of Cancer

Equator

Tropic of Capricorn

Air pressure

Air pressure is caused by the force of gravity in the Earth's atmosphere pulling air down towards the surface. In 1643, Galileo's pupil, Toricelli, invented the first instrument for measuring air pressure – the mercury barometer. Before weather maps were developed in the early 1800s, the barometer was the most important tool in weather forecasting. High pressure usually indicates fine, settled weather, and low pressure means cloudy, rainy weather. The French physicist, Jean de Borda (1733-1799), was the first to show that changes in air pressure are also related to wind speed. An aneroid (non-liquid) barometer measures the effect of air pressure on a chamber which has part of the air removed.

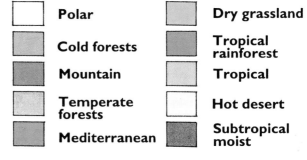

Aneroid barometer

	Polar		Dry grassland
	Cold forests		Tropical rainforest
	Mountain		Tropical
	Temperate forests		Hot desert
	Mediterranean		Subtropical moist

World climates

Climates depend on how near to the Equator a place is, how high it is above sea level and how far it is from the sea. World climates can be divided into the following categories:

Polar-Cold and snowy, strong winds
Cold forests-Short, summers and long, cold winters
Mountain-Cold and snowy high up
Temperate forests-Neither too hot nor too cold, rain all year
Mediterranean-Long, hot, dry summers and cool, wet winters
Dry grasslands-Hot, dry summers and cold, snowy winters
Tropical rainforest-Hot, rainy, humid, wet
Tropical-Hot all year, wet and dry seasons
Hot desert-Hot and dry, hardly any rain
Subtropical moist-Warm to hot summers, cool winters and moderate rain all year round.

EXTREME CLIMATES

The most important influence on the climate of a place is its distance from the Equator. Regions near the Equator are hot, and regions furthest from the Equator are cold. This is mainly because the Earth's surface is curved and the Sun's rays are more spread out at the Poles. Because every ray has to heat a larger area at the Poles, each has less heating power. These rays also have to travel further through the atmosphere and are weaker when they arrive.

Hot climates

Near the Equator, the Sun's rays are all at right angles to the Earth. They are concentrated on this small area, which makes these regions very hot. Tropical climates are damp and humid because the warm air rises and cools, forming clouds and rain. A huge variety of plants and animals thrive here. Desert climates are found a little closer to the Equator, in areas of high pressure. Here, cool air is sinking and becoming warmer and drier as it does so. Deserts cover about a seventh of the Earth's land surface. Plants like the cactus and the American mesquite tree have learned to store water in leaves, roots and stems, and to obtain water from deep beneath the surface. Deserts are spreading due to drier climates over the whole planet, the destruction of trees and overgrazing.

Animal adaptations

In the cold Polar regions, animals such as elephant seals and penguins have thick fur or feathers and layers of fat to keep warm. In the hot deserts, the desert hedgehog hides away in burrows during the heat of the day, coming out at dawn or dusk when it is cooler and moister. The addax can go for long periods without water. Its feet are broad and spreading to cope with desert sand. In the hot, humid tropical rainforests, animals have to shelter from the rain. The fur of the sloth hangs down so water runs off it easily.

Sloth

Desert hedgehog

Addax

Polar explorers

In 1908, Robert Peary was the first to reach the North Pole. Polar explorers Roald Amundsen and Captain Scott then raced to be the first to reach the South Pole in 1912. Amundsen won the race while Scott died tragically on the return journey. Modern Polar explorers, such as Sir Ranulf Fiennes and Geoff Somers use their journeys to find out more about the climate of these inhospitable regions.

Roald Amundsen

Robert Peary

Desert planet

Frank Herbert's (1920-1986) science fiction novel, *Dune* (1966), features the Planet Arrakis (also called Dune) – a world of great deserts and no open water. It is set 10,000 years into the future. Gigantic, mile-long sandworms produce a spice called melange, which allows travel through time and space, and gives people the ability to see into the future. The people of Arrakis eventually manage to change the climate of Dune and bring rain and water to the planet. Herbert wrote five other books in the Dune series.

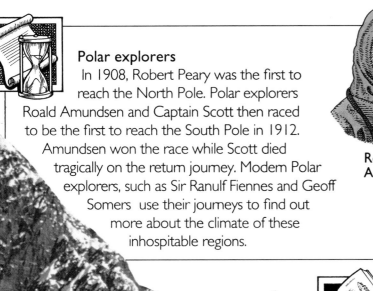

Penguin

Elephant seal

Cold climates

Climbing up a mountain is like making a journey from the Equator to the Poles. With increasing height, the air pressure drops because less air is pressing down from above. Temperatures fall about 1 degree centigrade for every 155 metres in height, and winds tend to be strong. Arctic regions are cold all year round although snow and ice may thaw during summer. Mosses, lichens and low shrubs grow in these areas, called the tundra (left). Each spring, tundras come to life and the ground is covered with bright flowers.

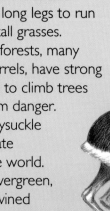

TEMPERATE CLIMATES

Mild, temperate climates are neither very hot nor very cold. They only cover about 7 per cent of the Earth's land surface, but over 40 per cent of the world's population live in them. Temperate climates are sandwiched between the cold Polar climates and the hot Equatorial climates. They have some rainfall all year round, but the temperature changes with the seasons, being warmer in summer and cooler in winter. In winter, it is usually too cold for plants to grow, so there is a dormant season when the plants rest.

Grasslands

Grasslands cover over a quarter of the Earth's land area. They are found where the climate is too dry and the soils too poor for most trees to survive. Temperate grasslands, such as the prairies of North America (right), the pampas of South America and the steppes of Central Asia have hot, dry summers and long, icy winters, with cold winds. The prairies are a large wheat-producing area.

Animal and plant adaptations

Grassland animals, like prairie dogs, maras and susliks, live in burrows where they are safe from fires and predators. Rheas have long legs to run fast and see over the tall grasses.

In woodlands and forests, many animals, such as squirrels, have strong legs and sharp claws to climb trees well and escape from danger.

Plants like the honeysuckle (left) grow in temperate woodlands around the world. They are hardy and evergreen, and often grow entwined around other plants and trees.

Prairie dog

Rhea

Mara

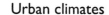

Urban climates

Towns and cities have an important influence on local climate. The buildings trap heat during the day and give off heat slowly at night, making a city up to 5°C (9°F) warmer than its surrounding areas. In summer, warm, moist air rising over skyscrapers may form clouds and rain. Wind speeds in cities are about 25 per cent slower than in the countryside. Pollution from vehicles and factories may be trapped near the ground, forming smog.

City pigeon

Forests and woods

In many temperate climates, there is enough rainfall for trees to grow. Deciduous trees, such as oak, beech, ash and maple, lose their leaves in winter when it is too cold for them to take up water from the frozen soil. In the taiga forests, which stretch right across the top of North America, Europe and Asia, the trees are mainly conifers, such as spruce and pine. They keep their needle-like leaves all year round and survive the long, cold winters.

Squirrel

Sea breezes

In summer, cool sea breezes blow off the sea in coastal areas. The land is warmer than the sea and as warm air rises over the land, cool air is drawn in from the sea to take its place. At night, the reverse happens. Cool air moves from the land to the sea because the land cools down faster than the sea. This is called a land breeze.

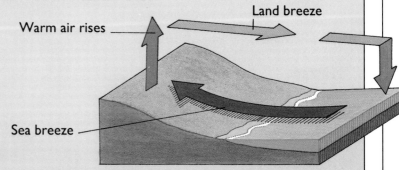

Warm air rises

Land breeze

Sea breeze

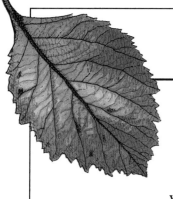

SEASONS

A season is a time of year with a particular kind of weather. Each season has a different effect on plant and animal life. Areas around the Poles have only two seasons – six months of summer, when it is light nearly all the time, and six months of winter, when it is dark most of the time. Places near the Equator have less defined seasons. Often there are only two, one wet and one dry. It is hot all year round, and the length of the day stays the same all year. Temperate regions between the Equator and the Poles, have four seasons – spring, summer, autumn and winter. The days are longer in summer and shorter in winter.

Life in autumn and winter

During these seasons, the weather may turn cooler, wetter and more windy. There is little food for animals to eat. Some gather stores of food in autumn to help them survive the winter. Plants also rest over the winter when it is too cold for them to grow and water in the soil is frozen. Areas closer to the Equator remain warm.

People in the north-east of Brazil (above) can still spend time on the beach, even in the winter.

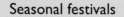

Winter in Canada (above) often brings snow.

Seasonal festivals

In the Northern Hemisphere, the Christian festival of Easter happens in springtime. Easter symbols, such as spring flowers and eggs, represent new life and the resurrection (or rising from the dead) of Jesus Christ. Some Hindu festivals are connected with the annual cycle of the seasons. Pongal or Sankranti marks the end of the south-east monsoon and the reaping of the harvest. Beautiful kilars (decorative designs) are traced on floors with moistened rice flour.

Life in spring and summer

Spring in temperate climates brings warmer weather and the days get longer. Day and night are almost the same length. The warmth and spring showers encourage plants and trees to grow and buds to burst open. Many animals have their young in spring so they will have time to grow strong enough to survive the cold autumn and winter seasons. Summer in the Mediterranean climate of Spain is very hot and dry. Olive trees (top left) are suited to this environment, and olive groves flourish.

Summer in South-East Asia (above) and parts of eastern Africa can be very wet when the monsoon rains arrive between April and July.

Spring bud

Hibernation and migration

To survive cold, hot or dry seasons, animals may move away or migrate to warmer, cooler or wetter places. The arctic tern migrates from one end of the world to the other and back again, covering about 40, 000 km a year. But other migrations, such as that of the wildebeest on the African grasslands, are over much shorter distances. Instead of moving away, other animals, such as doormice, stay put and go into a deep sleep in a safe place. This behaviour is called hibernation in a cold climate and aestivation in a hot climate.

Doormouse

Arctic tern

The four seasons

The Italian composer, violinist and conductor Antonio Vivaldi (1678-1741) wrote four famous violin concertos called "The Four Seasons" in about 1725. Each one consists of three pieces which convey the characteristics of each season. It is one of Vivaldi's best known compositions. Vivaldi wrote nearly 50 operas, church music and hundreds of concertos for almost every instrument known at the time.

LIVING WITH WEATHER

People have adapted to living in all sorts of climates. Because we can control our body temperature, choose clothes, design buildings and use different kinds of transport, we can survive in all sorts of climates and weather conditions. We can keep our homes warm with fires or central heating and cool them down with air-conditioning. In cold weather, we can wear layers of clothes to trap warm air near our bodies. In hot weather, we can wear loose clothes and use special creams to prevent sunburn.

Hot houses

Houses in hot climates are designed to provide shade. They have few partitions inside allowing cool air to move around freely. Some have shutters which people close during the day to keep the hot air out and open at night to let the cool air in. In tropical areas, houses are often built on stilts to let air circulate and to raise the living space above the flood level.

Camels are used for transport in deserts where vehicles easily overheat and get stuck in soft sand.

These desert houses are sunk into the ground to protect people from the intense heat by day and keep them warm at night. The white paint helps to reflect the heat away.

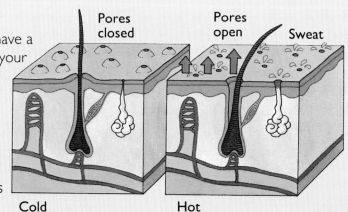

Goose-bumps galore

Next time you feel a bit chilly, have a look at your arm. You may see your hairs standing on end as the pores in your skin close tightly. This is just one technique your body uses to maintain its temperature of 37°C. If we get too cold, we shiver to generate heat from our muscles. If we get too hot, we sweat and send more blood to the skin's surface in order to lose heat.

Pores closed

Pores open

Sweat

Cold

Hot

Roman heating
In the stone houses in colder parts of their empire, the Romans constructed an underfloor heating system called a hypocaust. The floors of the rooms were raised up on piles of tiles or bricks. Hot air, heated by a furnace, circulated under the floors to heat the houses and baths. During the first century AD, channels were built into the walls and roofs so hot air could move around the whole building.

Stilt house

Hot air

Snow slides easily off these steep roofs (left).

Vehicles with ridged tyres provide extra grip for travelling on snow and ice. These Inuit Eskimos (below) use special snow-mobiles.

Cold houses
In cold climates, houses tend to be built of solid, heavy materials such as brick or stone so they won't blow over in strong winds or get washed away in the rain. Damp-proof courses (layers of brick or stone) stop buildings soaking up water from the ground. Thick walls help to keep the inside of the buildings warm and small windows prevent the heat from escaping. Double-glazing also helps to stop heat escaping and sloping roofs allow heavy falls of snow to slide to the ground.

SUN & TEMPERATURE

The Sun is the driving force behind the weather because it keeps the air in the atmosphere constantly on the move. The Sun's rays travel 150 million kilometres through space and warm the surface of the Earth, which then heats the air above it. This leads to the formation of winds and clouds (see pages 16-19). We measure temperature – how hot or cold the air is – with instruments called thermometers. Liquid mercury or alcohol inside a thermometer rises up a thin tube when it warms up, and sinks as it cools.

The Sun and Plants

The Sun's energy is captured by green plants and used to make their food. A green pigment in a plant's leaves, called chlorophyll, captures the sunlight. This is used to join carbon dioxide gas from the air with water from the soil to make sugars. The whole process is called photosynthesis. Animals cannot use the Sun's energy to make food like this, so they have to eat plants – or animals that have already eaten plants. All life on Earth depends on the energy in sunlight. This energy can be detected in the amount of carbon preserved in tree rings. Radioactive carbon 14 (C14) in the atmosphere seems to be linked to the activity of the Sun. When the Sun is less active, the Earth is colder. Very old trees, therefore, can tell us something about past climates on Earth.

Tree rings

Early experiments

In the mid-17th century, scientists at the Academy of Experiments in Italy developed a range of instruments to make the first planned observations of the weather. This engraving shows a selection of early thermometers. One early glass thermometer measured temperature by the rise and fall of coloured balls suspended in water. Simple mercury thermometers remain pretty much the same today as they were hundreds of years ago.

Solar power

The Sun is a source of safe, clean energy, but can be difficult to use because the weather changes day by day and with the seasons. The Sun's energy can be collected by satellites and transmitted back to Earth or collected on Earth using solar panels or mirrors that reflect the Sun onto a central heat collector. Solar power is used to generate electricity, to heat and cool buildings and to power objects from watches and calculators to cars and aeroplanes. It is a step towards a cleaner, safer planet.

90%

25%

Snow

Desert

Different surfaces (above) reflect different amounts of the Sun's heat. This is called the *albedo* of the surface. Forests reflect about 7% of the Sun's radiation, while snow and ice reflect most of it, keeping temperatures very low. Sandy deserts reflect only about 25% of the Sun's radiation, so daytime temperatures are extremely high as most of the Sun's heat is absorbed.

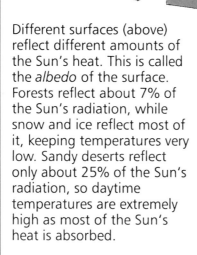

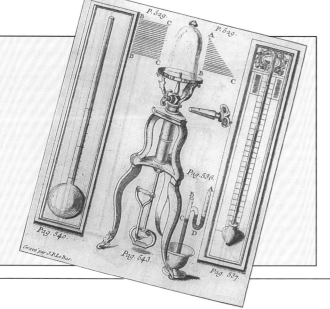

Skin colour

Human skin colour varies a great deal because it contains different amounts of a coloured substance – a pigment – called melanin. Melanin protects the skin from the Sun's harmful ultraviolet (UV) rays. Some people get a suntan in bright sunlight as the skin produces more melanin for protection. Sometimes melanin builds up in small spots, forming freckles. People who live in hot, sunny climates usually have a darker skin which contains a lot of melanin to protect them from the Sun. People with fair skins may burn easily in the Sun. Burning is very dangerous and may result in wrinkling or even skin cancer.

WINDY WEATHER

Winds are caused by air in the Earth's atmosphere moving from one place to another. The air moves from areas of high pressure to areas of low pressure – rather like letting air out of a balloon. The pressure differences are caused by the Sun heating up the air. Warm air is less dense or "lighter" than cool air, so it rises, creating low pressure. Cool air is more dense or "heavier". It sinks down and presses on the ground, creating high pressure.

Naming winds

The names of some winds tell you something about their characteristics. For instance, monsoon comes from "mausim", the Arab word for season, and describes winds in tropical areas that blow from the ocean during summer and towards the ocean in winter. In the horse latitudes, 30 degrees north and south of the Equator, there is little wind. This name may have come from the fact that many horses died on board ships held up by the lack of wind in the area. The chinook wind, which blows down the Rockies of North America in winter and early spring, was named by the first settlers who thought it came from the country of the Chinook Indians along the Columbia River.

Windmills

For over a thousand years, people have used the power of the wind to push round the sails of windmills. There are records of a farmer in Iran building a windmill as far back as AD 644. Nowadays, windmills are also used as a clean, efficient way to generate electricity. Some people think wind farms like this one in California, USA, spoil the landscape.

Blowing hot and cold

Over the whole globe, the world's winds move in a certain pattern. Hot air around the Equator heats up, rises and moves towards the Poles. As it cools it sinks back down, some returning to the Equator and some moving on to the Poles. Cold air at the Poles sinks and moves towards the Equator. Because the Earth is spinning round eastwards, winds blow from east to west (easterlies) as air moves towards the faster moving Equator. Winds blow from west to east (westerlies) as air moves away from the Equator towards the Poles. This is known as the Coriolis effect.

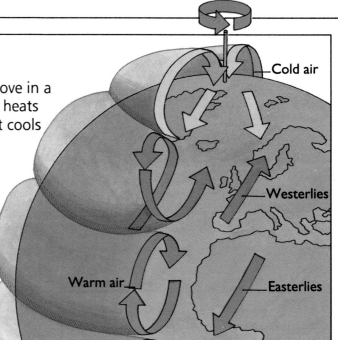

Cold air

Westerlies

Warm air

Easterlies

Sail away

In a yacht race, the competitors need to know about the wind's speed, the wind's direction and the force of the water pushing against the hull. Then they can work out how to position their sails. As the sails catch the wind, they bulge out on one side. The air pressure on the side that bulges is lower than on the other side, producng a pulling force which moves the yacht through the water.

Whistling wind

In his poem "The Rime of the Ancient Mariner", Samuel Taylor Coleridge (1772-1834) wrote of the feeling of being becalmed (without wind) in a ship in the doldrums. The doldrums are an area of calm winds or light breezes near the Equator.

"Day after day, day after day,
We stuck, nor breath nor motion:
As idle as a painted ship
Upon a painted ocean."

Try writing a poem of your own about the wind. How does the wind make you feel? What sort of noises does it make?

Boughing down

Trees near the coast and on windy hillsides often look as if they are bending over. In fact, as the wind usually blows more strongly in these areas, the tree grows more leaves on the sheltered side of the trunk, away from the wind. This makes the tree look lop-sided. Trees are anchored to the soil by long, widely-spread roots. These root systems are often as large as the crown of the tree, but hidden underground. They help to prevent the tree from becoming top-heavy and toppling over in a strong wind.

CLOUDS

Clouds are made up of millions of droplets of water or ice, which are so small and light they can float in the air. Clouds form when warm air rises. This happens when air is heated by the Sun or if it has to rise up over mountains or when cold air pushes it up from underneath. High in the sky, invisible water vapour in the air cools and turns into droplets of liquid water which gather together to make clouds. The shape, colour and height of clouds helps people to predict changes in the weather. Fog or mist are clouds that form down at ground level.

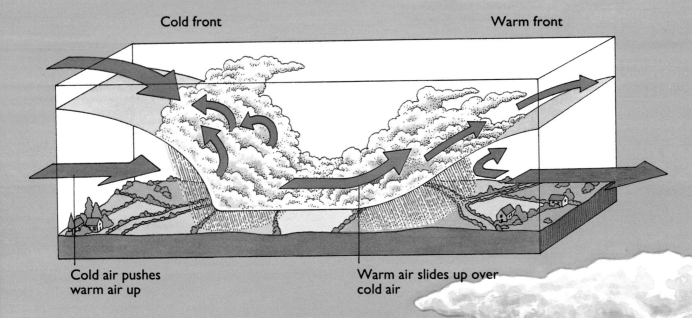

Cold front

Warm front

Cold air pushes warm air up

Warm air slides up over cold air

cumulonimbus – storm clouds. May rise to great heights while the bases are near the ground

Warm and cold fronts

Clouds often form where warm air meets cold air – this is called a weather front. The cold air may push up under the warm air, forcing it to rise rapidly. This is a cold front. The passing of the front brings colder weather behind. Or the warm air may slide slowly up over the cold air, forming a warm front. Warmer weather would follow this front. In both warm and cold fronts, warm air rises, cools and may form clouds. A weather front is a sign of change in the weather, with rain and sometimes storms as a result.

cumulus – heaped-up piles of fair-weather clouds

Heavens above

If someone asked you where heaven is, you'd probably point upwards towards the sky. Throughout history, heaven has been portrayed as a spiritual place above the clouds. This illustration by Gustave Dore (1832-1883) depicts a typical heavenly scene with winged angels supported by cotton-wool clouds. Films too, such as *Matter of Life and Death* (1946) show heaven as a timeless place with expansive floors of cloud.

cirrus – high ice clouds, often first to form along a weather front

altocumulus – sometimes referred to as a mackerel sky – sign of unsettled weather to come

Clouds in my coffee

When you look at the clouds, do they make you feel dreamy or sad or happy or hopeful? Clouds are quite often used to convey emotions in poetry and songs. Listen to the lyrics (words) of songs. How many can you think of that mention clouds, storms or rain?

stratocumulus - not as even in thickness as a stratus

stratus – rain or drizzle blanket clouds

Recycling the clouds

One of the ways clouds form is when the Sun heats water on the surface of the Earth. Some of the liquid water turns to water vapour and is absorbed into the air. This change from liquid water to water vapour is called evaporation. As the warm air, which is now full of moisture, rises up into the sky, it cools down. This makes the moisture turn back, or condense, into liquid water again, forming clouds. This is called the water cycle.

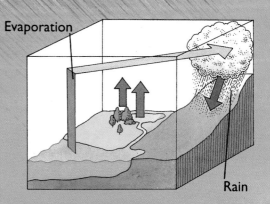

Evaporation

Rain

19

RAINY DAYS

Rain is formed in clouds when tiny water droplets bump into each other and join up to make bigger droplets, or when ice crystals warm up and melt into drops of water. When the drops get too heavy to float in the cloud, they fall to the ground as rain. Each raindrop is made up of about a million cloud droplets. Huge, dark cumulonimbus clouds give heavy, thundery showers, while low, flat stratus clouds produce a light drizzle. If the Sun shines through the rain, the raindrops separate the rays of light into the spectrum until all the different colours can be seen in the form of a rainbow.

Monsoon Rains

In tropical areas, such as India and South-East Asia, there are monsoon winds which blow in opposite directions in different seasons. This gives a rainy season and a dry season. Monsoons are like giant sea breezes (see page 9). For example, in India, most south-westerly winds blow off the Indian Ocean in early summer, producing torrential rain. The winds are drawn in from the ocean to replace hot air rising from the land. In winter, dry winds blow off the land towards the ocean, giving a dry season.

Monsoon rains fall in Simla, India

Bottle up

Bottle gardens hardly ever need watering because the plants use the same water over and over again. The water is recycled in the same way as the natural water cycle on Earth (see page 19). The plants take up water from the soil and give it off through their leaves in a process called transpiration. The water given off condenses into liquid water when it hits the cool sides of the bottle. This condensed water then runs back down into the soil where the plants can use it again.

Bottle garden

Raining cats and dogs

The strong air currents that cause heavy rain have been known to sweep up spiders, fish, maggots and even frogs. These creatures then fall with the rain some distance away. But no one has ever seen cats or dogs falling from the sky – as in the old saying. This expression may be based on the ancient Chinese spirits for rain and wind, which were sometimes illustrated as a cat and a dog.

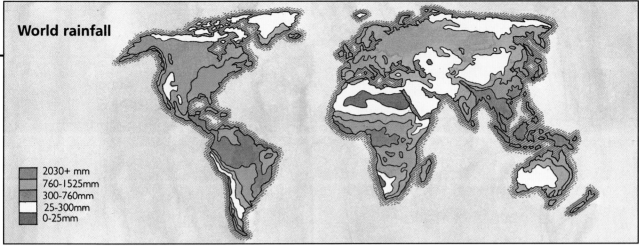

World rainfall

2030+ mm
760-1525mm
300-760mm
25-300mm
0-25mm

In heavy rain, some male chimpanzees charge about waving branches and making a lot of noise. No-one is sure why they do this wild "rain dance".

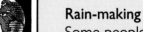

Rain-making

Some people believe that if rain doesn't fall, you can encourage it to start by singing and dancing. Very hot countries, like Birkina Fasso and Somalia in Africa, have very little rain, and often suffer from drought. Certain tribes of people have developed rain-making ceremonies which they believe will make the rain fall and the crops grow. On the Italian Island of Sicily during the feast of St John, men dance in an ancient ceremony to ask for rain. They hold reeds in their hands as they dance because reeds are a symbol of fresh water springs.

Waking deserts

Many desert plants survive the dry climate by resting as seeds buried in the sand and soil. When it rains, the seeds sprout quickly and the plants flower, making the desert come alive with a sea of colour. The plants rapidly produce seeds, which fall to the ground and wait in the soil for the next rains to arrive.

Acid rain

Acid rain occurs when gases like sulphur dioxide and nitrogen oxides from power stations and vehicle exhausts join up with water in the air to make rain more acidic. It also occurs as snow and sleet. Acid rain eats away at building stone and damages trees, rivers and lakes, killing fish, plants and other aquatic life. Acid rain occurs over areas of eastern North America, north-western and central Europe, parts of Asia and in other scattered locations around the world.

SNOW, ICE & HAIL

High in the sky, where the air temperature is below the freezing point of water, droplets of water in the clouds turn into ice crystals. More water then freezes on to the ice crystals, which grow bigger. As these crystals fall down through the cloud, they bump into other crystals and may form snowflakes. If the temperature near the ground is below freezing, snow falls from the clouds. But if is above freezing, the snowflakes melt and fall as rain or half-melted snow, called sleet. Icebergs are huge lumps of ice that break off the polar ice-caps.

Icy hazards

An avalanche can bury a village in seconds and smash trees as if they were matchsticks. It can happen when fresh snow falls on top of an icy layer on slopes. Avalanches can be triggered by a rise in temperature, a strong wind or even a loud noise. Icebergs are also a hazard. The passenger liner, Titanic, sank in April 1912 after hitting an iceberg. Lumps of ice that fall off icebergs are known as bergy bits, and even smaller lumps are called growlers.

Jumping hailstones

Hailstones are hard lumps of ice formed in cumulonimbus clouds when crystals of ice are thrown up and down by strong air currents. Ice builds up in layers around the ice crystals. Clear layers build up in the lower part of the cloud where it is warmer and the water freezes slowly; frosty ice layers form when the crystal is higher up in the cloud. By counting the layers, you can tell how many times a hailstone was tossed up and down inside a cloud.

This hailstone was tossed up and down five times in a cloud

Tracks in snow

Animal footprints in fresh snow provide clues to the variety of wildlife in an area. They show something of how the animals moved and what they were doing before you arrived on the scene. Can you guess which animals made these tracks? The answers are at the bottom of the page. You may find similar tracks in mud.

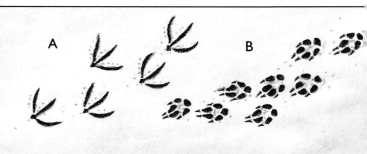

A   B

C. Domestic cat D. Mouse

Each snowflake is different although they all have six sides. The shape and size of snowflakes depends on the height and temperature at which they are formed and the amount of moisture in the cloud. In cold air, they are needle or rod-shaped; in warmer air, they are star or plate-shaped.

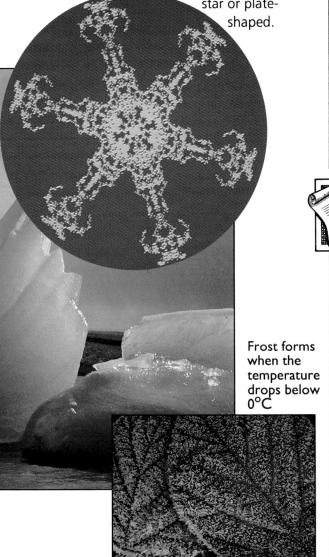

Frost forms when the temperature drops below 0°C

Jack Frost

The legendary Jack Frost is an elf-like figure who is supposed to leave his icy fingermarks on windowpanes. Beautiful patterns sometimes form on windows when water vapour turns directly to ice as it touches the freezing glass. The legend of Jack Frost probably comes from Norse Mythology, where Kari, god of the winds, had a son called Jokul (meaning icicle) or Frosti (meaning frost).

Jack Frost

The Ice Man

In September 1991, hikers in the Alps came across the body of a man who turned out to be over 5000 years old. He was preserved by being sealed in an airtight pocket beneath the ice of a glacier and the intense cold stopped his body decaying in the usual way. He was still wearing a boot stuffed with grass and his brain and internal organs were still intact. Bodies caught in glaciers are usually crushed and torn by the ice, so the fact that this body was preserved was pure chance. Scientists think the iceman froze to death after falling asleep. He may have been a mountain shepherd who had lost his weapons and was collecting material to make new ones.

The Ice Man

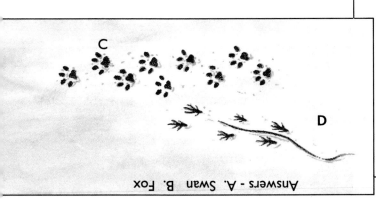

C

D

Answers - A. Swan B. Fox

EXTREME WEATHER

From thunderstorms and drought to hurricanes and tornadoes, extreme weather can bring disaster to many places around the world. In some parts of the world, extreme weather occurs each year as part of the changing seasons; for example, floods during the wet monsoon season in South-East Asia. But in other areas, extreme weather, such as hurricanes, are unusual. A lot of damage can be done when people are not prepared for the dangers. People who live in places where the weather and climate frequently become dangerous cannot always leave. Often they are very poor or they cannot move for political reasons.

Storms – wet and dry

The heavy rain that may fall during a storm often causes floods. If over 15mm of rain falls in three hours, this is called a "flash flood". Floods are more likely to occur after a dry spell when the ground is baked hard and water cannot drain away. Instead, it sweeps across the surface, washing away crops, houses and vehicles. Monsoon rains caused this flood in Bangladesh (below right). Dust storms occur where the ground has little or no protective vegetation. Strong, turbulent winds sweep up the dust, suspending it at a height of up to 300 metres. It may cover hundreds of kilometres like this one in Libya, North Africa (below).

Safe houses

In areas of heavy rain, such as Indonesia, houses are often built on stilts to keep them out of the way of the heavy monsoon rains. At certain times of year, water may reach floor level. Houses in stormy areas, like this one in Key West, Florida, may have storm shutters to protect the glass windows from fierce gusts of wind and flying debris. Roofs are also firmly fixed to the buildings. But there is little that can be done to safeguard homes from full-blown hurricanes or tornadoes. These whirling storms develop in warm, damp air when winds blow into each other from opposite directions. Hurricanes and tornadoes can demolish houses and uproot trees.

Gods of thunder

Thunderstorms were once thought to be started by powerful gods, such as Zeus the Greek god of the skies or Thor, the Norse god of thunder. These gods were supposed to forge thunderbolts and hurl them down from the sky. To prevent violent storms, people prayed to the gods or held ceremonies.

Zeus

A picture of Hurricane Gladys taken from space reveals the tell-tale swirling spiral of cloud. At the very centre of the hurricane is a small area of low pressure where the winds are light, the skies are clear and the air warm. This is called the "eye" of the storm. It is surrounded by winds spiralling around at up to 360kph. Hurricane Gloria hit Connecticut, USA (left) in 1985 causing these stormy seas and much damage to property.

Early forecasts

Robert FitzRoy, captain of HMS Beagle during Darwin's voyage from 1831-1836, was also a pioneer of weather forecasting. He collected information on the weather around the coasts of Britain and drew up the first weather charts. He also introduced a system of storm warning signals. His forecasts did not always turn out to be right, however, in the same way that modern forecasts often prove to be unreliable.

Fitzroy

Storm shutters

Stilt houses

CHANGING THE WEATHER

The weather and climate of the Earth has changed many times during the 4,600 million years since the Earth began. There have been cold periods, called Ice Ages, lasting thousands of years, with warm periods in-between. The last Ice Age was 10,000 years ago. Ice Ages may be caused by a change in the Earth's orbit around the Sun or a change in the tilt of the Earth on its axis. Nowadays, many scientists believe we are causing the weather to change by polluting the atmosphere with chemicals, such as CFCs, and the gases from power stations and vehicles.

Holes over the Poles

About 25 km above the Earth is a layer of gas called ozone, a form of oxygen. This ozone layer stops too many of the Sun's ultra-violet rays from reaching the Earth. They can cause skin cancer and stop plants growing. Holes have been discovered in the ozone layer all over the planet, but especially over the Antarctic (below) and Arctic because of the special weather conditions there. Gases called CFCs are probably to blame. They are used in some aerosols, refrigerators, air-conditioning, and some polystyrene.

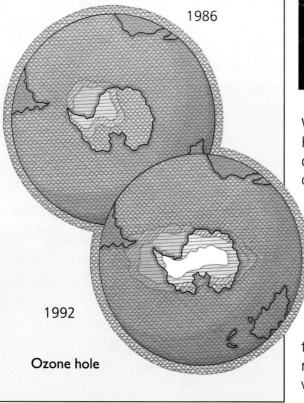

1986

1992

Ozone hole

When a big volcano, such as Mount St. Helen's (above), erupts, it throws huge clouds of dust high into the atmosphere, blocking out the Sun and making the Earth cooler. This volcanic dust may stay in the air for many years. When the Earth first formed, thousands of volcanoes probably covered the planet. They threw out gases and dust into the atmosphere which probably contributed to a dramatic cooling of the Earth's climate. Agricultural and industrial activity also cause large amounts of particles to be discharged into the air. These particles may have effects similar to those caused by volcanic dust, cooling the Earth's climate.

Frost fairs

From 1600-1800, the weather was much colder than it is today and the River Thames in England froze over regularly. Frost fairs were held on the ice. In 1683-84, the River Thames was completely frozen for two months. The last frost fair on the Thames was in 1814, when the ice was thick enough to take the weight of an elephant.

Forests and climate

Many forests in the Himalayas have been cut down for firewood and timber, so there are now fewer trees to soak up heavy rainfall. The water pours down the mountains, causing floods as far away as the Ganges delta. As trees make food in a process called photosynthesis (see page 14), they take in a gas called carbon dioxide from the air and give off oxygen. When trees are cut down, the levels of carbon dioxide in the atmosphere build up. This gas traps some of the heat that rises from the Earth when it is warmed by the Sun, and stops it escaping into space. This is called the greenhouse effect. Extra carbon dioxide is making the atmosphere warm up in a process called global warming.

Animals from the past

Dinosaurs, like Triceratops (left), ruled the Earth for over 150 million years when the climate was much warmer and more humid. They died out about 65 million years ago, possibly as a result of the climate getting colder. This may have been caused by a huge asteroid from outer space hitting the Earth and sending up dust to blot out the Sun's heat. During the last ice ages, woolly mammoths survived because of their thick, shaggy coats. They died out when the climate became warmer.

Woolly mammoth

FORECASTING

Predicting or forecasting the weather is an important part of planning our everyday lives. People need to know what clothes to wear, if they can travel safely and efficiently, and if it's possible for them to work outdoors. Farmers, builders and people who run transport systems, such as railways or airlines are always monitoring weather information. Nowadays, thousands of measurements on land, sea and in the air are taken every day to record things like temperature, wind speed and rainfall. Computers process the information so that weather maps, called synoptic charts, can be drawn up.

Signs of the times
Sayings and rhymes help us to remember signs of change in the weather. "Red sky at night is a shepherd's delight" means that the weather is likely to be fine after a red sunset. In North America, February 2nd is groundhog day. If the groundhog wakes up, sees his own shadow and goes back to sleep, the saying goes that winter will last another six weeks.

Pine-cones are a good way to tell if the weather is going to be wet or dry. When they dry out, the cones open up like these; when it's damp, the cone absorbs the water and the scales close up tightly.

Weather watching
The information used in weather forecasting comes from a variety of sources. These vary from weather stations on land and ships, and weather buoys at sea to planes and weather balloons high in the sky and satellites out in space. Weather satellites take photographs of cloud patterns (which are clues to changes in the weather) and also infrared pictures, which show the temperature of the Earth. Some weather satellites stay in one place, while others circle around the Earth.

The weather plane, "Snoopy"

Special weather planes record the weather at various levels in the atmosphere by taking 3D pictures and monitoring temperature and humidity.

Weather hairs

It is important for weather forecasters to know how much water there is in the air. This is called its humidity. A human hair can be used to measure humidity since it stretches in moist air and shrinks in dry air. This principle is how weather houses work. When the air is humid, a hair inside the house stretches, and the man comes out of the door. When the air dries out, the hair shrinks, pulling the man back inside. The woman then swings out, forecasting fine, dry weather.

A weather house

A weather house is a kind of hair hygrometer. Hygrometers can be made small enough to fit into a pocket, and are very useful for walkers to take with them on a hike in the countryside. Some early hygrometers used a paper strip to move a needle. The paper shrank or stretched as it responded to the dampness in the air. Humidity affects our comfort and health. When the temperature and humidity are high people feel uncomfortable and "sticky" because their perspiration doesn't evaporate. People may use air-conditioners to take the water vapour out of the air.

The weather satellite, above, is called Meteosat I. It was launched in 1977 and operated until 1985. It recorded images of the whole Earth over Africa and Europe, gathering weather data. Satellites carry television cameras that take pictures of the Earth. These show the pattern of clouds and large areas of snow and ice. Satellites also beam picture signals to stations on the ground.

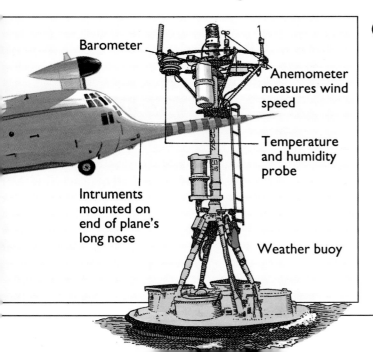

Barometer

Anemometer measures wind speed

Temperature and humidity probe

Instruments mounted on end of plane's long nose

Weather buoy

Weather shaping history

During the Second World War, forecasting the weather was important for a number of battles. The D-day landing of the allies in Normandy, France in 1944 was postponed because of bad weather. And in 1941-42, the Germans were beaten back from Moscow and Stalingrad partly by the fierce resistance, and partly by the bitter winter weather. Weather has also influenced history in other ways. For example, an unexpected southerly gale drove Captain Cook towards what was then the unknown coast of Australia.

WONDERFUL WEATHER FACTS

Dinosaurs lived when the Earth was warmer and died out about 65 million years ago.

An American called "snowflake" Bentley spend 50 years studying thousands of snowflakes, but never found two the same.

Wettest place in the world is Hawaii – 11,684 mm of rain falls per year, and it rains almost every day.

Highest recorded temperature was in Libya on edge of Sahara desert. It was 58°C (136.4°F) in the shade. The highest wind speed recorded was 450kph (280mph) during a tornado in Texas, USA.

The Empire state building in New York does not have lightning conductors. It is possible for it to be struck 500 times a year, and it has been struck 48 times in one day.

Avalanches can move at up top 320 kph (200mph) down a mountainside.

Driest place in the world is in Chile – in the Atacam desert it did not rain at all for 400 years, until 1971.

Jet streams are very strong winds that save about 1 hr on the flying time from New York to London.

Antarctica has little rainfall - only about 70 mm (2.8in) of snow at the South Pole.

Each sqm of the Earth's surface has about 10,000 kg of air above it.

The Moon has no atmosphere to retain heat, so temperature drops to -140°C when it is in the shade.

The Earth intercepts only one part in 2 billion of the energy given off by the Sun, and 35 per cent of this is reflected back into space.

The amount of moisture in the atmosphere at any one time is equivalent to only about 25mm (1in) of rainfall when it is spread over the whole surface of the Earth.

Scientists predict that Earth will warm by 2-4°C (1.8-3.6°F) by 2030 unless greenhouse gases cut, The Statue of Liberty will be underwater.

Each year there are about 16 million thunderstorms in the world; that's about 1,800 at any one time.

The biggest hailstone fell in Bangladesh. It weighed 1.02kg, about same as a bag of sugar.

Air in a tornado can whirl at up to 800kph (500mph), Many buildings explode as the pressure inside them cannot fall as quickly as the drop in outside pressure.

Early 19th century balloonists, James Glaisher and Robert Coxwell, risked their lives to find out about atmosphere.

GLOSSARY

Acid rain Rain which is more acidic than normal because of pollution.

Air pressure The weight of all the air in the Earth's atmosphere pressing down on the surface of the planet.

Albedo The amount of the Sun's radiation reflected from a surface.

Atmosphere The protective blanket of gases that surrounds the Earth.

Barometer An instrument which measures air pressure.

CFCs (chlorofluorocarbons) Chemicals which are mainly responsible for destroying the ozone layer.

Climate The average weather in a place over a long period of time.

Condensation The process by which a gas or a vapour changes into a liquid as it cools down.

Cyclone A hurricane in the Indian Ocean and the seas north of Australia.

Evaporation The process by which a liquid is heated to become a vapour or gas.

Front A region where warm air meets cold air, often causing rain.

Frost Ice formed when water vapour in the air condenses on cold surfaces and freezes.

Global warming An increase in the Earth's temperature which may be caused by carbon dioxide and other greenhouse gases stopping the Earth's heat from escaping.

Hail Pellets of ice which form in a cumulonimbus storm cloud as air currents toss frozen raindrops up and down.

Humidity The amount of water vapour (moisture) in the air.

Hurricane A violent tropical storm

Hygrometer An instrument for measuring humidity.

Meteorology The scientific study of the atmosphere and the weather.

Monsoon A wind which blows from different directions at different times of year, causing a wet season and a dry season.

Ozone layer A layer of the gas called ozone in the Earth's atmosphere which absorbs 90 per cent of the harmful ultra-violet radiation from the Sun.

Temperature How hot or cold something is. It is measured in degrees farenheit (^{o}F) or centigrade (^{o}C).

Thermometer An instrument used to measure temperature.

Thunder The noise caused by the air expanding quickly and exploding when it is heated by lightning.

Tornado A tall, funnel-shaped whirlwind of air which may extend from the bottom of a cumulonimbus cloud to the ground.

Troposphere The lowest layer of the atmosphere, above the Earth's surface, where the weather happens.

Warm-blooded The ability of some animals (such as humans) to keep their body temperature the same, no matter how hot or cold it is around them.

Wind Air moving from areas of high pressure to areas of low pressure. Also effected by the Earth spinning on its axis.

INDEX

acid rain 21, 31
adaptation 6, 8, 12
air pressure 5, 7, 16, 17, 31
albedo 15, 31
animals 6, 8, 10, 11, 14,
  21, 22, 27
atmosphere 4, 31
auroras 4
avalanches 22

barometers 5, 31
body temperature 12, 31
breezes 9, 20

CFCs 26, 31
changing the weather 26-7
chinook wind 16
climate 4, 31
clouds 18-19, 20
condensation 19, 20, 31
Coriolis effect 17
cumulonimbus clouds 18,
  20, 22

deserts 6, 12, 15, 21, 30
doldrums 17
dust storms 24

Equator 6, 10, 17
evaporation 19, 31

feathers 6
floods 24, 27
fog and mist 18
forests 6, 9, 15, 27
frost 23, 27, 31

global warming 27, 30, 31
grasslands 8
greenhouse effect 27
growlers 22

hailstones 22, 30, 31
horse latitudes 16
house design 12-13, 24
humidity 29, 30, 31
hurricanes 24, 25, 31
hygrometers 29, 31

ice 13, 15, 22, 23, 27
Ice Ages 26, 27
icebergs 22

jet streams 30

lightning 30

meteorology 14, 25, 28-9
  31
monsoons 11, 16, 20, 24
  31

ozone layer 26, 31

photosynthesis 14, 27
pine-cones 28
plants 6, 7, 8, 10, 14, 20,
  21
Poles 6, 7, 10, 17, 26, 30
pollution 9, 21, 26

rain 9, 18, 20-1, 30
rainbows 20

seasonal festivals 10
seasons 4, 10-11
skin colour 15
sleet 21, 22
smog 9
snow 13, 15, 21, 22
snowflakes 22, 23, 30
solar power 15
storms 18, 24
stratus clouds 19, 20
Sun 4, 6, 14, 15, 16, 19,
  26, 30

temperate climates 8-9, 10
  11
temperatures 7, 9, 14, 15
  30, 31
thermometers 14, 31
thunderstorms 25, 30, 31
tornadoes 24, 30, 31
tropical climates 6, 12, 20
troposphere 4, 31

urban climates 9

volcanoes 26

water cycle 19, 20
weather 4
weather forecasting 25, 28-
  9
weather fronts 18, 31
weather satellites 28, 29
wind speeds 5, 9, 17, 30
windmills 16
winds 16-17, 20, 24, 31

Photographic Credits:
Abbreviations: t-top, m-middle, b-bottom, l-left, r-right
Cover t, 4t, 23tl, 24mr, 26m, 29t: Science Photo Library; cover b, title p, 2t, 5, 9r, 10-11, 12t, 14t & b, 16ml, 17, 20b, 26t, 27, 28br: Roger Vlitos; 2b, 6t, 8 both, 13, 15r, 18, 20m, 22t, 23bl, 24ml, 24b, 25t, 28t: Eye Ubiquitous; 3, 4b, 19, 23tr, 29b: Mary Evans Picture Library; 6-7, 91, 11t, 12m, 14m, 16mr & b, 22m, 24t, 25b: Spectrum Colour Library; 7b, 10m, 11m, 23br: Frank Spooner Pictures; 10t: Suzanne Melia; 11b, 151: Hulton Deutsch; 28bl: Dan Brooks.